MW01097773

Harcdcover: 9798372299955

First paperback edition January 2023.

Exploring the Cosmos with James Webb

STEEVEN CADEL

Exploring the Cosmos with James Webb

STEEVEN CADEL

Who is James WEBB?

NASA *Administrator, February* 14, 1961-*October* 7, 1968

James Edwin Webb was the second administrator of the National Aeronautics and Space Administration, formally established on October 1, 1958, under the National Aeronautics and Space Act of 1958.

Born on October 7, 1906, in Tally Ho, North Carolina, he was the son of John Frederick and Sarah Gorham Webb. His father was superintendent of schools in Granville County for 26 years. In 1938 he married Patsy Aiken Douglas and they had two children: Sarah Gorham, born on February 27, 1945, and James Edwin Jr., born on March 5, 1947.

Mr. Webb was educated at the University of North Carolina, where he received an A.B. in education in 1928. He became a second lieutenant in the Marine Corps and served as a pilot on active duty from 1930-1932. He also studied law at George Washington University from 1934-1936 and was admitted to the Bar of the District of Columbia in 1936.

He enjoyed a long career in public service, coming to Washington in 1932 and serving as secretary to Congressman Edward W. Pou, 4th North Carolina District, Chair of the House Rules Committee, until 1934. He then served as assistant in the office of O. Max Gardner, attorney and former governor of North Carolina, in Washington, D.C., between 1934 and 1936. In 1936 Mr. Webb became personnel director, secretary-treasurer and later vice president of the Sperry Gyroscope Company in Brooklyn, New York, before re-entering the U.S. Marine Corps in 1944 for World War II.

After World War II, Mr. Webb returned to Washington and served as executive assistant to O. Max Gardner, by then Under Secretary of the Treasury, before being named as director of the Bureau of the Budget in the Executive Office of the President, a position he held until 1949. President Harry S. Truman then asked Mr. Webb to serve as Under Secretary of State, U.S. Department of State. When the Truman administration ended early in 1953, Mr. Webb left Washington for a position in the Kerr-McGee Oil Corp. in Oklahoma City, OK.

James Webb returned to Washington on February 14, 1961, when he accepted the position of administrator of NASA. Under his direction the agency undertook one of the most impressive projects in history, the goal of landing an American on the Moon before the end of the decade through the execution of Project Apollo.

For seven years after President Kennedy's May 25, 1961, lunar landing announcement, through October 1968, James Webb politicked, coaxed, cajoled, and maneuvered for NASA in Washington. As a longtime Washington insider he was a master at bureaucratic politics. In the end, through a variety of methods Administrator Webb built a seamless web of political liaisons that brought continued support for and resources to accomplish the Apollo Moon landing on the schedule President Kennedy had announced.

Mr. Webb was in the leadership of NASA when tragedy struck the Apollo program. On January 27, 1967, Apollo-Saturn (AS) 204, was on the launch pad at Kennedy Space Center, Florida, moving through simulation tests when a flash fire killed the three astronauts aboard--"Gus" Grissom, Edward White, and Roger Chaffee.
Shock gripped NASA and the nation during the days that followed. James Webb told the media at the time, "We've always known that something like this was going to happen soon or later. . . . who would have thought that the first tragedy would be on the ground?" As the nation mourned, Webb went to President Lyndon Johnson and asked that NASA be allowed to handle the accident investigation and direct the recovery from the accident. He promised to be truthful in assessing blame and pledged to assign it to himself and NASA management as appropriate. The agency set out to discover the details of the tragedy, to correct problems, and to get back on schedule.
Mr. Webb reported these findings to various Congressional committees and took a personal grilling at nearly every meeting. While the ordeal was personally taxing, whether by happenstance or design Webb deflected much of the backlash over the fire from both NASA as an agency and from the Johnson administration. While he was personally tarred with the disaster, the space agency's image and popular support was largely undamaged. He left NASA in October 1968, just as Apollo was nearing a successful completion.
After retiring from NASA, Mr. Webb remained in Washington, D.C., serving on several advisory boards, including as a regent of the Smithsonian Institution. He died on March 27, 1992 and is buried in Arlington National Cemetary.

Credit: NASA

Webb Uncovers Star Formation in Cluster's Dusty Ribbons

NGC 346, one of the most dynamic star-forming regions in nearby galaxies, is full of mystery. Now, though, it is less mysterious thanks to new findings from the NASA/ESA/CSA James Webb Space Telescope.

NCG 346 is located in the Small Magellanic Cloud (SMC), a dwarf galaxy close to our Milky Way. The SMC contains lower concentrations of elements heavier than hydrogen or helium, which astronomers call metals, than seen in the Milky Way. Since dust grains in space are composed mostly of metals, scientists expected that there would only be small amounts of dust, and that it would be hard to detect. But new data from Webb reveals just the opposite.

Astronomers probed this region because the conditions and amount of metals within the SMC resemble those seen in galaxies billions of years ago, during an era in the Universe's history known as 'cosmic noon,' when star formation was at its peak. Some 2 to 3 billion years after the Big Bang, galaxies were forming stars at a furious rate. The fireworks of star formation happening then still shape the galaxies we see around us today.

"A galaxy during cosmic noon wouldn't have one NGC 346, as the Small Magellanic Cloud does; it would have thousands", said Margaret Meixner, an astronomer at the Universities Space Research Association and principal investigator of the research team. "But even if NGC 346 is now the one and only massive cluster furiously forming stars in its galaxy, it offers us a great opportunity to probe the conditions that were in place at cosmic noon."

By observing protostars still in the process of forming, researchers can learn if the star formation process in the SMC is different from what we observe in our own Milky Way. Previous infrared studies of NGC 346 have focused on protostars heavier than about five to eight times the mass of our Sun. "With Webb, we can probe down to lighter-weight protostars, as small as one tenth of our Sun, to see if their formation process is affected by the lower metal content," said Olivia Jones of the United Kingdom Astronomy Technology Centre, at the Royal Observatory Edinburgh, a co-investigator on the program.

As stars form, they gather gas and dust, which can look like ribbons in Webb imagery, from the surrounding molecular cloud. The material collects into an accretion disc that feeds the central protostar. Astronomers have detected gas around protostars within NGC 346, but Webb's near-infrared observations mark the first time they have also detected dust in these discs.

"We're seeing the building blocks, not only of stars, but also potentially of planets," said Guido De Marchi of the European Space Agency, a co-investigator on the research team. "And since the Small Magellanic Cloud has a similar environment to that of galaxies during cosmic noon, it's possible that rocky planets could have formed earlier in the history of the Universe than we might have thought."

Credit: NASA, ESA, CSA, STScI, A. Pagan (STScI)

Dusty Debris Disk Around AU Mic

The NASA/ESA/CSA James Webb Space Telescope has imaged the inner workings of a dusty disk surrounding a nearby red dwarf star. These observations represent the first time the previously known disk has been imaged at these infrared wavelengths of light. They also provide clues to the composition of the disk.

These two images are of the dusty debris disk around AU Mic, a red dwarf star located 32 light-years away in the southern constellation Microscopium. Scientists used Webb's Near-Infrared Camera (NIRCam) to study AU Mic. NIRCam's coronagraph, which blocked the intense light of the central star, allowed the team to study the region very close to the star. The location of the star, which is masked out, is marked by a white, graphical representation at the center of each image. The region blocked by the coronagraph is shown by a dashed circle.

Webb provided images at 3.56 microns (top, blue) and 4.44 microns (bottom, red). The team found that the disk was brighter at the shorter or "bluer" wavelength, likely meaning that it contains a lot of fine dust that is more efficient at scattering shorter wavelengths of light.

The NIRCam images allowed the researchers to trace the disk, which spans a diameter of 60 astronomical units (9 billion kilometers), as close to the star as 5 astronomical units (740 million kilometers) – the equivalent of Jupiter's orbit in our solar system. The images were more detailed and brighter than the team expected, and scientists were able to image the disk closer to the star than expected.

While detecting the disk is significant, the team's ultimate goal is to search for giant planets in wide orbits, similar to Jupiter, Saturn, or the ice giants of our solar system. Such worlds are very difficult to detect around distant stars using either the transit or radial velocity methods.

These results are being presented in a press conference at the 241st meeting of the American Astronomical Society. The observations were obtained as part of Webb's Guaranteed Time program 1184.

[Image Description: The visual shows two bright lines, representing the dusty debris disc around the red dwarf star AU Mic. The glowing line on top is blue, representing 3.56 microns and appears brighter, and the glowing line on bottom is red, representing 4.44 microns.]

Credit:NASA, ESA, CSA, and K. Lawson (Goddard Space Flight Center), A. Pagan (STScI)

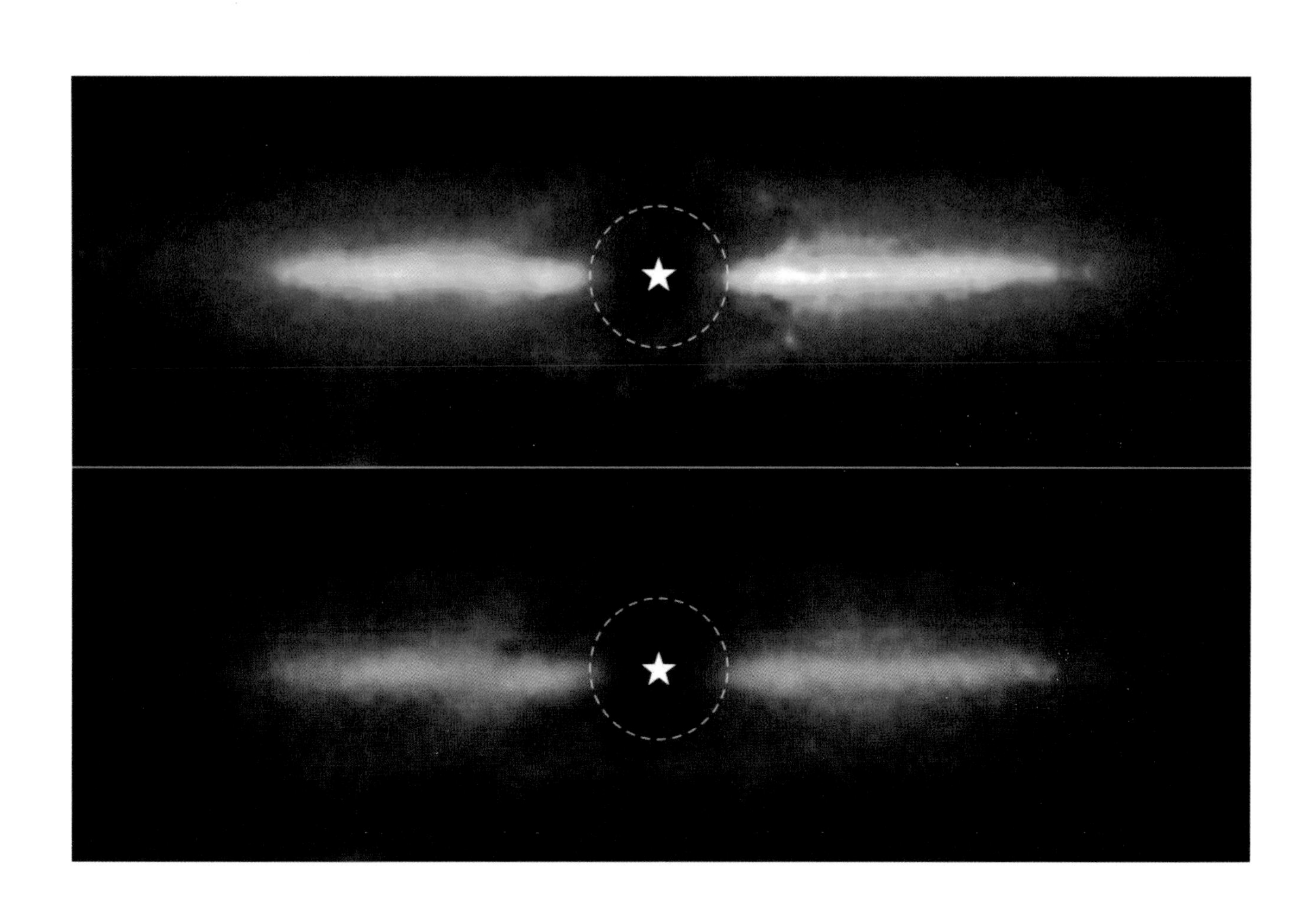

Webb Reveals Complex Galactic Structures

This image from the NASA/ESA/CSA James Webb Space Telescope shows IC 5332, a spiral galaxy, in unprecedented detail thanks to observations from the Mid-InfraRed Instrument (MIRI). Its symmetrical spiral arms, which appear so clearly in Hubble's ultraviolet and visible-light image of IC 5332, are revealed as a complex web of gas, emitting infrared light at a variety of temperatures.

Capturing light at these wavelengths requires very specialised instruments kept at very cold temperatures, and MIRI performs spectacularly at the task.

MIRI – the Mid-InfraRed Instrument
Webb has a suite of four powerful instruments that will investigate the cosmos. They are located in the Integrated Science Instrument Module, behind the primary mirror.

MIRI is the only instrument on the telescope that is capable of operating at mid-infrared wavelengths. It will support the whole range of Webb's science goals, from observing our own Solar System and other planetary systems, to studying the early Universe.
MIRI is a versatile instrument offering a wide set of modes: imaging, coronagraphy and different flavours of spectroscopy. To observe the cosmos in the mid-infrared, MIRI must be kept more than 30 degrees cooler than the other instruments in the Webb observatory. This is accomplished by the use of an innovative cooling system known as a cryocooler, which will act as an extra refrigerator for the MIRI instrument.

MIRI will provide imaging, coronagraphy and spectroscopy over the 5 µm- 28 µm wavelength range. It will operate at −266°C (compared with −233°C for the rest of the observatory). This is barely seven degrees above absolute zero, the lowest temperature possible according to the laws of physics.

MIRI was developed as a partnership between Europe and the USA: the main partners are ESA, a consortium of nationally funded European institutes, the Jet Propulsion Laboratory (JPL) and GSFC. The instrument was nationally funded by the European Consortium under the auspices of the European Space Agency. The principal investigator leading the MIRI European Consortium is Gillian Wright (UK Astronomy Technology Centre) and the MIRI American science lead is George Rieke (University of Arizona).

Credit:ESA/Webb, NASA & CSA, J. Lee and the PHANGS-JWST and PHANGS-HST Teams

Webb Takes a Stunning, Star-Filled Portrait of the Pillars of Creation

The Pillars of Creation are set off in a kaleidoscope of colour in the NASA/ESA/CSA James Webb Space Telescope's near-infrared-light view. The pillars look like arches and spires rising out of a desert landscape, but are filled with semi-transparent gas and dust, and ever changing. This is a region where young stars are forming – or have barely burst from their dusty cocoons as they continue to form.

Protostars are the scene-stealers in this Near-Infrared Camera (NIRCam) image. These are the bright red orbs that sometimes appear with eight diffraction spikes. When knots with sufficient mass form within the pillars, they begin to collapse under their own gravity, slowly heat up, and eventually begin shining brightly.

Along the edges of the pillars are wavy lines that look like lava. These are ejections from stars that are still forming. Young stars periodically shoot out jets that can interact within clouds of material, like these thick pillars of gas and dust. This sometimes also results in bow shocks, which can form wavy patterns like a boat does as it moves through water. These young stars are estimated to be only a few hundred thousand years old, and will continue to form for millions of years.

Although it may appear that near-infrared light has allowed Webb to "pierce through" the background to reveal great cosmic distances beyond the pillars, the interstellar medium stands in the way, like a drawn curtain.

This is also the reason why there are no distant galaxies in this view. This translucent layer of gas blocks our view of the deeper universe. Plus, dust is lit up by the collective light from the packed "party" of stars that have burst free from the pillars. It's like standing in a well-lit room looking out a window – the interior light reflects on the pane, obscuring the scene outside and, in turn, illuminating the activity at the party inside.

Webb's new view of the Pillars of Creation will help researchers revamp models of star formation. By identifying far more precise star populations, along with the quantities of gas and dust in the region, they will begin to build a clearer understanding of how stars form and burst out of these clouds over millions of years.

The Pillars of Creation is a small region within the vast Eagle Nebula, which lies 6,500 light-years away.

Webb's NIRCam was built by a team at the University of Arizona and Lockheed Martin's Advanced Technology Center.

Credit:NASA, ESA, CSA, STScI; J. DePasquale, A. Koekemoer, A. Pagan (STScI).

Webb Inspects the Heart of the Phantom Galaxy

This image from the NASA/ESA/CSA James Webb Space Telescope shows the heart of M74, otherwise known as the Phantom Galaxy. Webb's sharp vision has revealed delicate filaments of gas and dust in the grandiose spiral arms which wind outwards from the centre of this image. A lack of gas in the nuclear region also provides an unobscured view of the nuclear star cluster at the galaxy's centre. M74 is a particular class of spiral galaxy known as a 'grand design spiral', meaning that its spiral arms are prominent and well-defined, unlike the patchy and ragged structure seen in some spiral galaxies.

The Phantom Galaxy is around 32 million light-years away from Earth in the constellation Pisces, and lies almost face-on to Earth. This, coupled with its well-defined spiral arms, makes it a favourite target for astronomers studying the origin and structure of galactic spirals.

Webb gazed into M74 with its Mid-InfraRed Instrument (MIRI) in order to learn more about the earliest phases of star formation in the local Universe. These observations are part of a larger effort to chart 19 nearby star-forming galaxies in the infrared by the international PHANGS collaboration. Those galaxies have already been observed using the NASA/ESA Hubble Space Telescope and ground-based observatories. The addition of crystal-clear Webb observations at longer wavelengths will allow astronomers to pinpoint star-forming regions in the galaxies, accurately measure the masses and ages of star clusters, and gain insights into the nature of the small grains of dust drifting in interstellar space.

Hubble observations of M74 have revealed particularly bright areas of star formation known as HII regions. Hubble's sharp vision at ultraviolet and visible wavelengths complements Webb's unparalleled sensitivity at infrared wavelengths, as do observations from ground-based radio telescopes such as the Atacama Large Millimeter/submillimeter Array, ALMA. By combining data from telescopes operating across the electromagnetic spectrum, scientists can gain greater insight into astronomical objects than by using a single observatory – even one as powerful as Webb!

Credit:ESA/Webb, NASA & CSA, J. Lee and the PHANGS-JWST Team.
Acknowledgement: J. Schmidt

Pillars of Creation

The NASA/ESA/CSA James Webb Space Telescope's mid-infrared view of the Pillars of Creation strikes a chilling tone. Thousands of stars that exist in this region disappear from view – and seemingly endless layers of gas and dust become the centrepiece.

The detection of dust by Webb's Mid-Infrared Instrument (MIRI) is extremely important – dust is a major ingredient for star formation. Many stars are actively forming in these dense blue-grey pillars. When knots of gas and dust with sufficient mass form in these regions, they begin to collapse under their own gravitational attraction, slowly heat up, and eventually form new stars.

Although the stars appear to be missing, they aren't. Stars typically do not emit much mid-infrared light. Instead, they are easiest to detect in ultraviolet, visible, and near-infrared light. In this MIRI view, two types of stars can be identified. The stars at the end of the thick, dusty pillars have recently eroded most of the more distant material surrounding them but they can be seen in mid-infrared light because they are still surrounded by cloaks of dust. In contrast, blue tones indicate stars that are older and have shed most of their gas and dust.

Mid-infrared light also details dense regions of gas and dust. The red region toward the top, which forms a delicate V shape, is where the dust is both diffuse and cooler. And although it may seem like the scene clears toward the bottom left of this view, the darkest grey areas are where densest and coolest regions of dust lie. Notice that there are many fewer stars and no background galaxies popping into view.

Webb's mid-infrared data will help researchers determine exactly how much dust is in this region – and what it's made of. These details will make models of the Pillars of Creation far more precise. Over time, we will begin to understand more clearly how stars form and burst out of these dusty clouds over millions of years.

Credit:NASA, ESA, CSA, STScI, J. DePasquale (STScI), A. Pagan (STScI)

Carina Nebula Jets

Scientists taking a "deep dive" into one of the iconic first images from the NASA/ESA/CSA James Webb Space Telescope have discovered dozens of energetic jets and outflows from young stars previously hidden by dust clouds. The discovery marks the beginning of a new era of investigating how stars like our Sun form, and how the radiation from nearby massive stars might affect the development of planets.

Dozens of previously hidden jets and outflows from young stars are revealed in this new image from Webb's Near-Infrared Camera (NIRCam). This image separates out several wavelengths of light from the First Image revealed on 12 July 2022, which highlights molecular hydrogen, a vital ingredient for star formation.

The Cosmic Cliffs, a region at the edge of a gigantic, gaseous cavity within the star cluster NGC 3324, has long intrigued astronomers as a hotbed for star formation. While well-studied by the NASA/ESA Hubble Space Telescope, many details of star formation in NGC 3324 remain hidden at visible-light wavelengths. Webb is perfectly primed to tease out these long-sought-after details since it is built to detect jets and outflows seen only in the infrared at high resolution. Webb's capabilities also allow researchers to track the movement of other features previously captured by Hubble.

Recently, by analyzing data from a specific wavelength of infrared light (4.7 microns), astronomers discovered two dozen previously unknown outflows from extremely young stars revealed by molecular hydrogen. Webb's observations uncovered a gallery of objects ranging from small fountains to burbling behemoths that extend light-years from the forming stars. Many of these protostars are poised to become low mass stars, like our Sun.

Molecular hydrogen is a vital ingredient for making new stars and an excellent tracer of the early stages of their formation. As young stars gather material from the gas and dust that surround them, most also eject a fraction of that material back out again from their polar regions in jets and outflows. These jets then act like a snowplow, bulldozing into the surrounding environment. Visible in Webb's observations is the molecular hydrogen getting swept up and excited by these jets.

Previous observations of jets and outflows looked mostly at nearby regions and more evolved objects that are already detectable in the visual wavelengths seen by Hubble. The unparalleled sensitivity of Webb allows observations of more distant regions, while its infrared optimization probes into the dust-sampling younger stages. Together this provides astronomers with an unprecedented view into environments that resemble the birthplace of our solar system.

In analyzing the new Webb observations, astronomers are also gaining insights into how active these star-forming regions are, even in a relatively short time span. By comparing the position of previously known outflows in this region caught by Webb, to archival data by Hubble from 16 years ago, the scientists were able to track the speed and direction in which the jets are moving.

In this image, red, green, and blue were assigned to Webb's NIRCam data at 4.7, 4.44, and 1.87 microns (F470N, F444W, and F187N filters, respectively).

Tarantula Nebula (NIRCam Image)

In this mosaic image stretching 340 light-years across, Webb's Near-Infrared Camera (NIRCam) displays the Tarantula Nebula star-forming region in a new light, including tens of thousands of never-before-seen young stars that were previously shrouded in cosmic dust. The most active region appears to sparkle with massive young stars, appearing pale blue. Scattered among them are still-embedded stars, appearing red, yet to emerge from the dusty cocoon of the nebula. NIRCam is able to detect these dust-enshrouded stars thanks to its unprecedented resolution at near-infrared wavelengths.

To the upper left of the cluster of young stars, and the top of the nebula's cavity, an older star prominently displays NIRCam's distinctive eight diffraction spikes, an artefact of the telescope's structure. Following the top central spike of this star upward, it almost points to a distinctive bubble in the cloud. Young stars still surrounded by dusty material are blowing this bubble, beginning to carve out their own cavity. Astronomers used two of Webb's spectrographs to take a closer look at this region and determine the chemical makeup of the star and its surrounding gas. This spectral information will tell astronomers about the age of the nebula and how many generations of star birth it has seen.

Farther from the core region of hot young stars, cooler gas takes on a rust colour, telling astronomers that the nebula is rich with complex hydrocarbons. This dense gas is the material that will form future stars. As winds from the massive stars sweep away gas and dust, some of it will pile up and, with gravity's help, form new stars.

Credit:ESA/Webb, NASA & CSA, L. Armus, A. S. Evans

A Wreath of Star Formation in NGC 7469

This image is dominated by NGC 7469, *a luminous, face-on spiral galaxy approximately* 90 000 *light-years in diameter that lies roughly* 220 *million light-years from* Earth *in the constellation* Pegasus. *Its companion galaxy* IC 5283 *is partly visible in the lower left portion of this image.*

This spiral galaxy has recently been studied as part of the Great Observatories All-*sky* LIRGs *Survey* (GOALS) *Early Release Science program with the* NASA/ESA/CSA *James Webb Space Telescope, which aims to study the physics of star formation, black hole growth, and feedback in four nearby, merging luminous infrared galaxies. Other galaxies studied as part of the survey include previous* ESA/Webb Pictures *of the* Month II ZW 096 *and* IC 1623.

NGC 7469 *is home to an active galactic nucleus* (AGN), *which is an extremely bright central region that is dominated by the light emitted by dust and gas as it falls into the galaxy's central black hole. This galaxy provides astronomers with the unique opportunity to study the relationship between* AGNs *and starburst activity because this particular object hosts an* AGN *that is surrounded by a starburst ring at a distance of a mere* 1500 *light-years.* While NGC 7469 *is one of the best studied* AGNs *in the sky, the compact nature of this system and the presence of a great deal of dust have made it difficult for scientists to achieve both the resolution and sensitivity needed to study this relationship in the infrared. Now, with* Webb, *astronomers can explore the galaxy's starburst ring, the central* AGN, *and the gas and dust in between.*

Using Webb's MIRI, NIRCam *and* NIRspec *instruments to obtain images and spectra of* NGC 7469 *in unprecedented detail, the* GOALS *team has uncovered a number of details about the object.* This *includes very young star-forming clusters never seen before, as well as pockets of very warm, turbulent molecular gas, and direct evidence for the destruction of small dust grains within a few hundred light-years of the nucleus – proving that the* AGN *is impacting the surrounding interstellar medium.* Furthermore, *highly ionised, diffuse atomic gas seems to be exiting the nucleus at roughly* 6.4 *million kilometres per hour – part of a galactic outflow that had previously been identified, but is now revealed in stunning detail with* Webb. With *analysis of the rich* Webb *datasets still underway, additional secrets of this local* AGN *and starburst laboratory are sure to be revealed.*

A prominent feature of this image is the striking six-pointed star that perfectly aligns with the heart of NGC 7469. *Unlike the galaxy, this is not a real celestial object, but an imaging artifact known as a diffraction spike, caused by the bright, unresolved* AGN. *Diffraction spikes are patterns produced as light bends around the sharp edges of a telescope.* Webb's *primary mirror is composed of hexagonal segments that each contain edges for light to diffract against, giving six bright spikes. There are also two shorter, fainter spikes, which are created by diffraction from the vertical strut that helps support* Webb's *secondary mirror.*

Credit:ESA/Webb, NASA & CSA, L. Armus, A. S. Evans

Southern Ring Nebula's Spokes

This is an image of the Southern Ring Nebula (NGC 3132), captured by the NASA/ESA/CSA James Webb Space Telescope's Near-Infrared Camera (NIRCam) and Mid-Infrared Instrument (MIRI). The image combines near- and mid-infrared light from three filters.

Webb's image traces the star's scattered outflows that have reached farther into the cosmos. Most of the molecular gas that lies outside the band of cooler gas is also cold. It is also far clumpier, consisting of dense knots of molecular gas that form a halo around the central stars.

Credit:NASA, ESA, CSA, STScI, O. De Marco (Macquarie University), J. DePasquale (STScI)

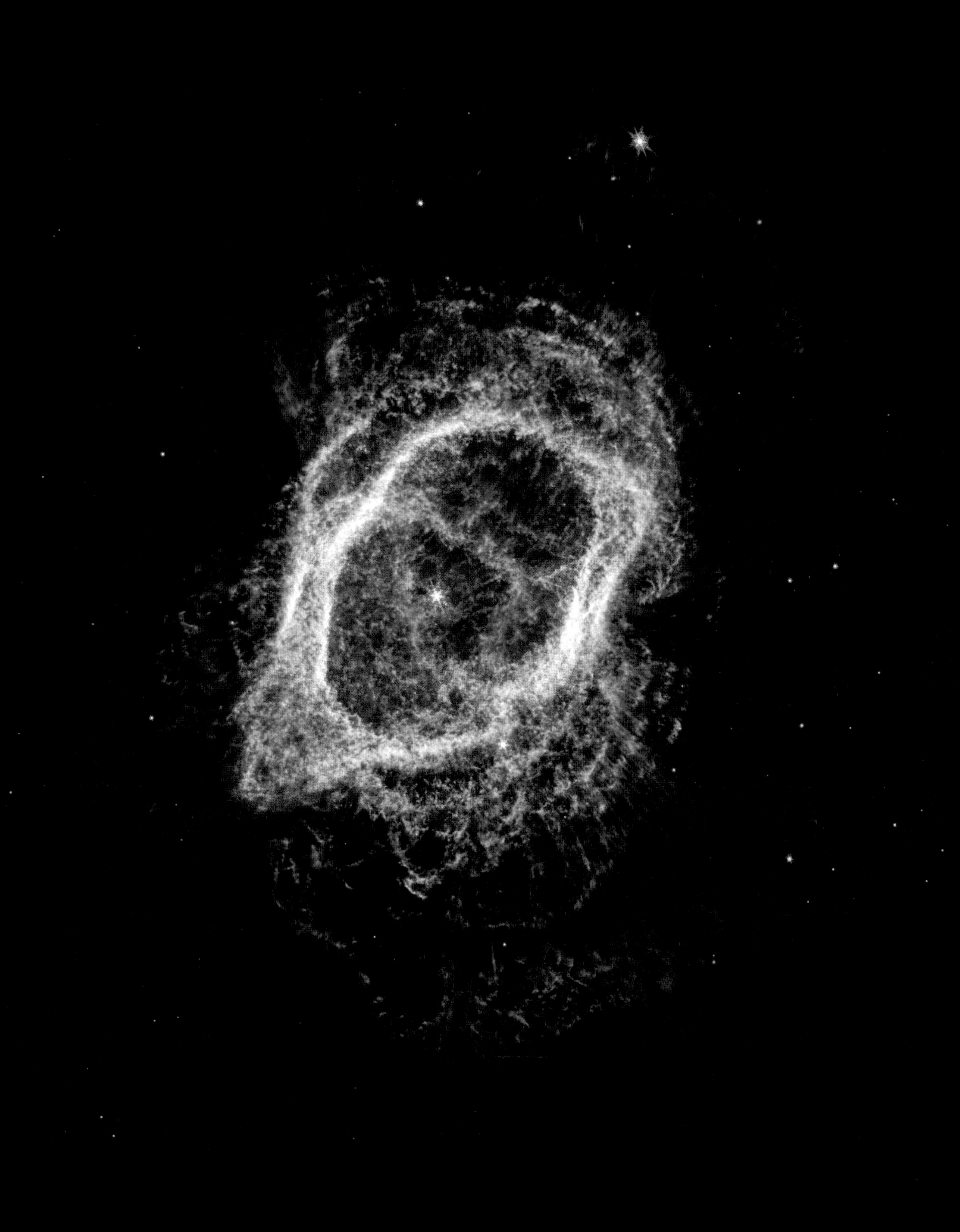

Webb Finds Distant Galaxies Seen Behind Pandora's Cluster

Two of the most distant galaxies seen to date are captured in these Webb pictures of the outer regions of the giant galaxy cluster Abell 2744. The galaxies are not inside the cluster, but many billions of light-years behind it.

The galaxy featured in the image at the top centre is extracted from the image on the left. It existed only 450 million years after the Big Bang.

The galaxy featured in the image at the bottom centre is extracted from the image on the right. It existed 350 million years after the Big Bang.

Both galaxies are seen really close in time to the Big Bang which occurred 13.8 billion years ago.

These galaxies are tiny compared to our Milky Way, being just a few percent of its size, even the unexpectedly elongated galaxy showcased in the top centre image.

Credit:NASA, ESA, CSA, T. Treu (UCLA)

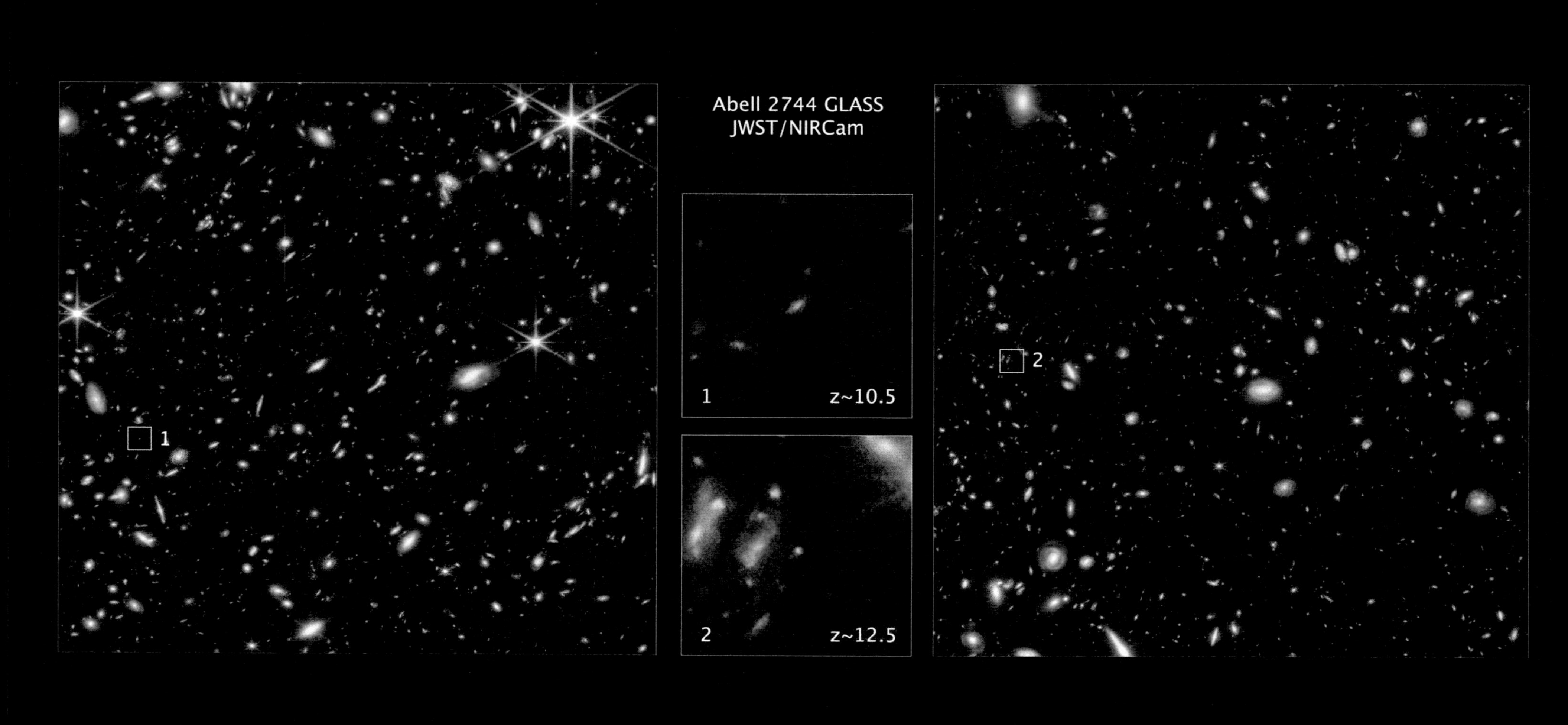
Abell 2744 GLASS
JWST/NIRCam
1
2
1
z~10.5
2
z~12.5

Protostar L1527

The protostar L1527, shown in this image from the NASA/ESA/CSA James Webb Space Telescope, is embedded within a cloud of material that is feeding its growth. Material ejected from the star has cleared out cavities above and below it, whose boundaries glow orange and blue in this infrared view. The upper central region displays bubble-like shapes due to stellar 'burps,' or sporadic ejections. Webb also detects filaments made of molecular hydrogen that has been shocked by past stellar ejections. Intriguingly, the edges of the cavities at upper left and lower right appear straight, while the boundaries at upper right and lower left are curved. The region at lower right appears blue, as there's less dust between it and Webb than the orange regions above it.

Credit:NASA, ESA, CSA, and STScI, J. DePasquale (STScI)

Webb's View Around the Extremely Red Quasar SDSS J165202.64+172852.3

The quasar SDSS J165202.64+172852.3 is highlighted in an image from the NASA/ESA Hubble Space Telescope in visible and near-infrared on the left. The images in the centre and on the right present new observations from the NASA/ESA/CSA James Webb Space Telescope in multiple wavelengths to demonstrate the distribution of gas around the object.

The quasar is an "extremely red" quasar that exists in the very early Universe, 11.5 billion years ago.

The image in the centre is composed of four narrow-band images made from the Webb NIRSpec instrument's integral-field spectroscopy mode. All the four narrow-band images show extremely red-shifted emissions from doubly ionised oxygen which has an emission line around 500nm in visible light; before it was shifted to infrared light.

The panels on the right present the four narrow-band images separately. Each colour illustrates the relative speed of ionised oxygen gas across the cluster. The redder the colour the faster gas is moving away from our line of sight with the quasar, while the bluer the colour the faster it's moving away from the quasar toward us. The colour green indicates that the gas is steady in our line of sight in comparison to the quasar.

The blue and yellow panels reveal the bi-conical outflow from the quasar, with the orange panel showing the gas moving faster from us, which is extended towards the lower right, as well as highlighting a companion galaxy on the upper left of the frame.

Credit:ESA/Webb, NASA & CSA, D. Wylezalek, A. Vayner & the Q3D Team, N. Zakamska

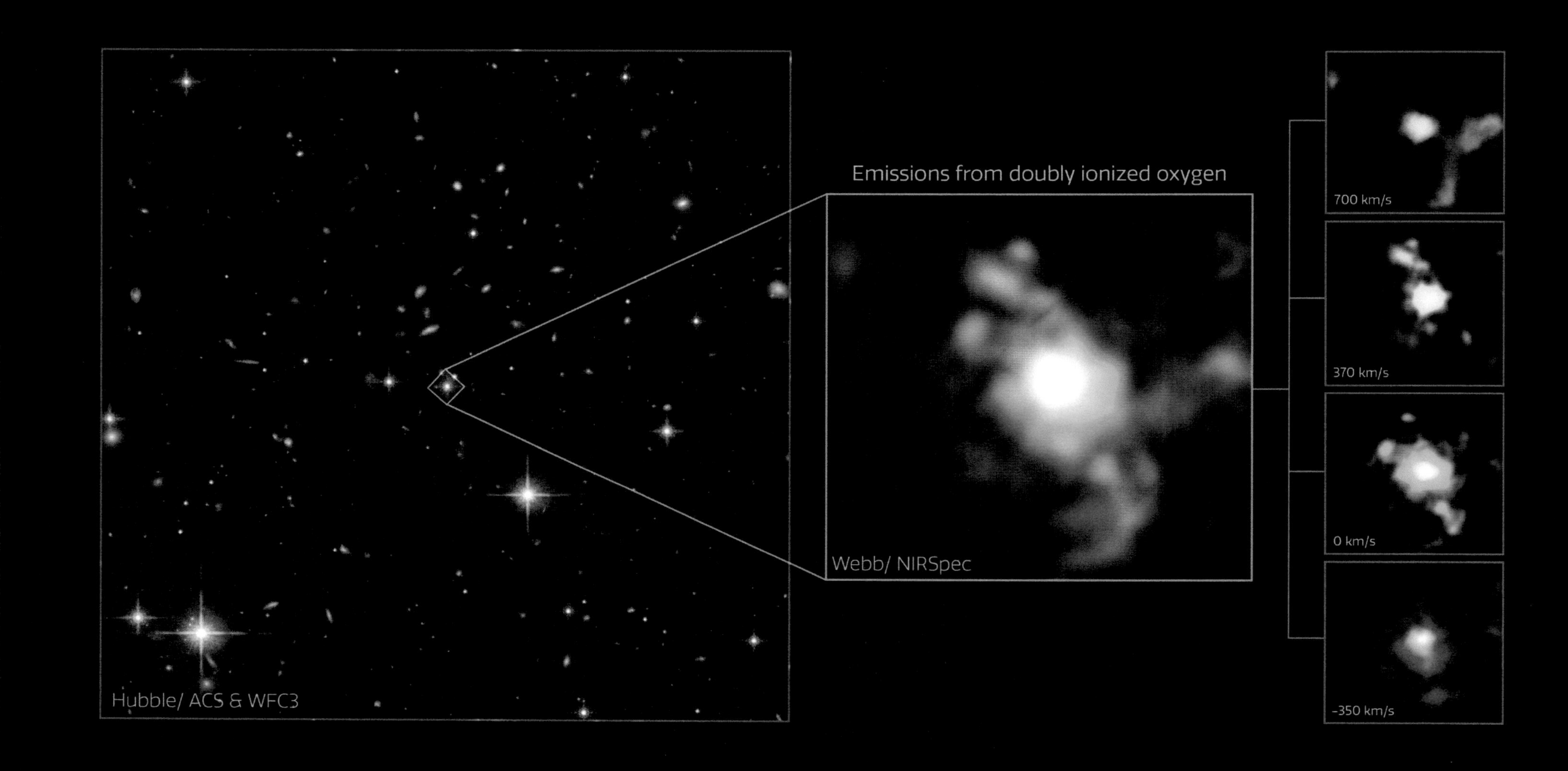
Hubble/ ACS & WFC3
Emissions from doubly ionized oxygen
Webb/ NIRSpec
700 km/s
370 km/s
0 km/s
-350 km/s

Galactic Get-Together

A merging galaxy pair cavort in this image captured by the NASA/ESA/CSA James Webb Space Telescope. This pair of galaxies, known to astronomers as II ZW 96, is roughly 500 million light-years from Earth and lies in the constellation Delphinus, close to the celestial equator. As well as the wild swirl of the merging galaxies, a menagerie of background galaxies are dotted throughout the image.

The two galaxies are in the process of merging and as a result have a chaotic, disturbed shape. The bright cores of the two galaxies are connected by bright tendrils of star-forming regions, and the spiral arms of the lower galaxy have been twisted out of shape by the gravitational perturbation of the galaxy merger. It is these star-forming regions that made II ZW 96 such a tempting target for Webb; the galaxy pair is particularly bright at infrared wavelengths thanks to the presence of the star formation.

This observation is from a collection of Webb measurements delving into the details of galactic evolution, in particular in nearby Luminous Infrared Galaxies such as II ZW 96. These galaxies, as the name suggests, are particularly bright at infrared wavelengths, with luminosities more than 100 billion times that of the Sun. An international team of astronomers proposed a study of complex galactic ecosystems – including the merging galaxies in II ZW 96 – to put Webb through its paces soon after the telescope was commissioned. Their chosen targets have already been observed with ground-based telescopes and the NASA/ESA Hubble Space Telescope, which will provide astronomers with insights into Webb's ability to unravel the details of complex galactic environments.

Webb captured this merging galaxy pair with a pair of its cutting-edge instruments; NIRCam – the Near-InfraRed Camera – and MIRI, the Mid-InfraRed Instrument.

Credit:ESA/Webb, NASA & CSA, L. Armus, A. Evans

Webb Explores a Pair of Merging Galaxies

This image from the NASA/ESA/CSA James Webb Space Telescope depicts IC 1623, an entwined pair of interacting galaxies which lies around 270 million light-years from Earth in the constellation Cetus. The two galaxies in IC 1623 are plunging headlong into one another in a process known as a galaxy merger. Their collision has ignited a frenzied spate of star formation known as a starburst, creating new stars at a rate more than twenty times that of the Milky Way galaxy.

This interacting galaxy system is particularly bright at infrared wavelengths, making it a perfect proving ground for Webb's ability to study luminous galaxies. A team of astronomers captured IC 1623 across the infrared portions of the electromagnetic spectrum using a trio of Webb's cutting-edge scientific instruments: MIRI, NIRSpec, and NIRCam. In so doing, they provided an abundance of data that will allow the astronomical community at large to fully explore how Webb's unprecedented capabilities will help to unravel the complex interactions in galactic ecosystems. These observations are also accompanied by data from other observatories, including the NASA/ESA Hubble Space Telescope, and will help set the stage for future observations of galactic systems with Webb.

The merger of these two galaxies has long been of interest to astronomers, and has previously been imaged by Hubble and by other space telescopes. The ongoing, extreme starburst causes intense infrared emission, and the merging galaxies may well be in the process of forming a supermassive black hole. A thick band of dust has blocked these valuable insights from the view of telescopes like Hubble. However, Webb's infrared sensitivity and its impressive resolution at those wavelengths allows it to see past the dust and has resulted in the spectacular image above, a combination of MIRI and NIRCam imagery.

The luminous core of the galaxy merger turns out to be both very bright and highly compact, so much so that Webb's diffraction spikes appear atop the galaxy in this image. The 8-pronged, snowflake-like diffraction spikes are created by the interaction of starlight with the physical structure of the telescope. The spiky quality of Webb's observations is particularly noticeable in images containing bright stars, such as Webb's first deep field image.

Credit:ESA/Webb, NASA & CSA, L. Armus & A. Evans
Acknowledgement: R. Colombari

Multi-Observatory Views of M74

New images of the Phantom Galaxy, M74, showcase the power of space observatories working together in multiple wavelengths. On the left, the NASA/ESA Hubble Space Telescope's view of the galaxy ranges from the older, redder stars towards the centre, to younger and bluer stars in its spiral arms, to the most active stellar formation in the red bubbles of H II regions. On the right, the NASA/ESA/CSA James Webb Space Telescope's image is strikingly different, instead highlighting the masses of gas and dust within the galaxy's arms, and the dense cluster of stars at its core. The combined image in the centre merges these two for a truly unique look at this "grand design" spiral galaxy.

Scientists combine data from telescopes operating across the electromagnetic spectrum to truly understand astronomical objects. In this way, data from Hubble and Webb compliment each other to provide a comprehensive view of the spectacular M74 galaxy.

Credit:ESA/Webb, NASA & CSA, J. Lee and the PHANGS-JWST Team; ESA/Hubble & NASA, R. Chandar
Acknowledgement: J. Schmidt

Hubble / Optical
Hubble & Webb
Webb / Infrared

Webb Explores a Pair of Merging Galaxies

Here, the Webb Picture of merging galaxies IC 1623 A and B is juxtaposed with a new image from the NASA/ESA Hubble Space Telescope. In the Webb MIRI image, the bright core, heated gas and dust, and young star forming regions are all visible. The Hubble and Webb NIRCAM images show the galaxies distorted spiral arms, while MIRI reveals the faint ghostly glow of interstellar dust.

Credit:ESA/Webb, NASA & CSA, L. Armus & A. Evans
Acknowledgement: R. Colombari

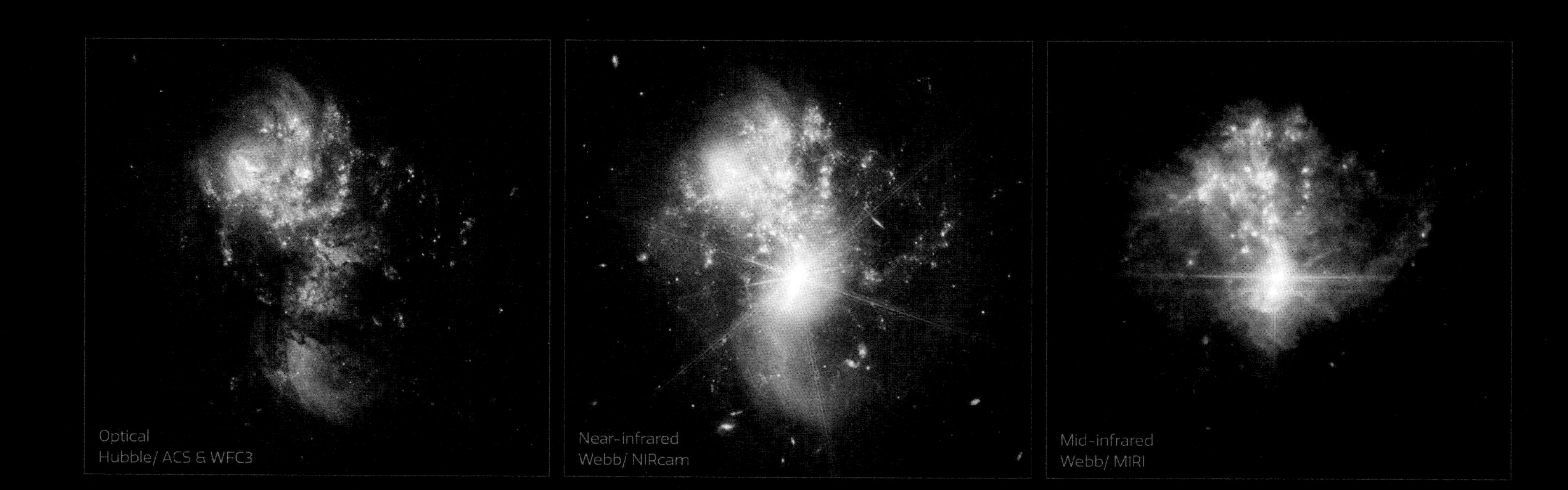
Optical
Hubble/ ACS & WFC3
Near-infrared
Webb/ NIRcam
Mid-infrared
Webb/ MIRI

Galaxy Pair VV 191 (Webb and Hubble Composite Image)

By combining data from the NASA/ESA/CSA James Webb Space Telescope and the NASA/ESA Hubble Space Telescope, researchers were able to trace light that was emitted by the large white elliptical galaxy at left through the spiral galaxy at right and identify the effects of interstellar dust in the spiral galaxy. This image of galaxy pair VV 191 includes near-infrared light from Webb, and ultraviolet and visible light from Hubble.

Webb's near-infrared data also show us the galaxy's longer, extremely dusty spiral arms in far more detail, giving them an appearance of overlapping with the central bulge of the bright white elliptical galaxy on the left. Although the two foreground galaxies are relatively close astronomically speaking, they are not actively interacting.

Don't overlook the background scenery! Like many Webb images, this image of VV 191 shows many galaxies that lie great distances away. For example, two patchy spirals to the upper left of the elliptical galaxy have similar apparent sizes, but show up in very different colors. One is likely very dusty and the other very far away, but researchers need to obtain data known as spectra to determine which is which.

Credit:NASA, ESA, CSA, Rogier Windhorst (ASU), William Keel (University of Alabama), Stuart Wyithe (University of Melbourne), JWST PEARLS Team, Alyssa Pagan (STScI)

North Ecliptic Pole Time Domain Field

The NASA/ESA/CSA James Webb Space Telescope has captured one of the first medium-deep wide-field images of the cosmos, featuring a region of the sky known as the North Ecliptic Pole. The image, which accompanies a paper published in the Astronomical Journal, is from the Prime Extragalactic Areas for Reionization and Lensing Science (PEARLS) GTO program.

"Medium-deep" refers to the faintest objects that can be seen in this image, which are about 29th magnitude (1 billion times fainter than what can be seen with the unaided eye), while "wide-field" refers to the total area that will be covered by the program, about one-twelfth the area of the full moon. The image is composed of eight different colors of near-infrared light captured by Webb's Near-Infrared Camera (NIRCam), augmented with three colors of ultraviolet and visible light from the NASA/ESA Hubble Space Telescope. This beautiful color image unveils in unprecedented detail and to exquisite depth a universe full of galaxies to the furthest reaches, many of which were previously unseen by Hubble or the largest ground-based telescopes, as well as an assortment of stars within our own Milky Way galaxy. The NIRCam observations will be combined with spectra obtained with Webb's Near-Infrared Imager and Slitless Spectrograph (NIRISS), allowing the team to search for faint objects with spectral emission lines, which can be used to estimate their distances more accurately.

A swath of sky measuring 2% of the area covered by the full moon was imaged here with NIRCam instrument in eight filters and with Hubble's Advanced Camera for Surveys (ACS) and Wide-Field Camera 3 (WFC3) in three filters that together span the 0.25 – 5-micron wavelength range. This image represents a portion of the full PEARLS field, which will be about four times larger. Thousands of galaxies over an enormous range in distance and time are seen in exquisite detail, many for the first time. Light from the most distant galaxies has traveled almost 13.5 billion years to reach us. Because this image is a combination of multiple exposures, some stars show additional diffraction spikes.

This representative-color image was created using Hubble filters F275W (purple), F435W (blue), and F606W (blue); and Webb filters F090W (cyan), F115W (green), F150W (green), F200W (green), F277W (yellow), F356W (yellow), F410M (orange), and F444W (red).

Credit:NASA, ESA, CSA, A. Pagan (STScI) & R. Jansen (ASU)

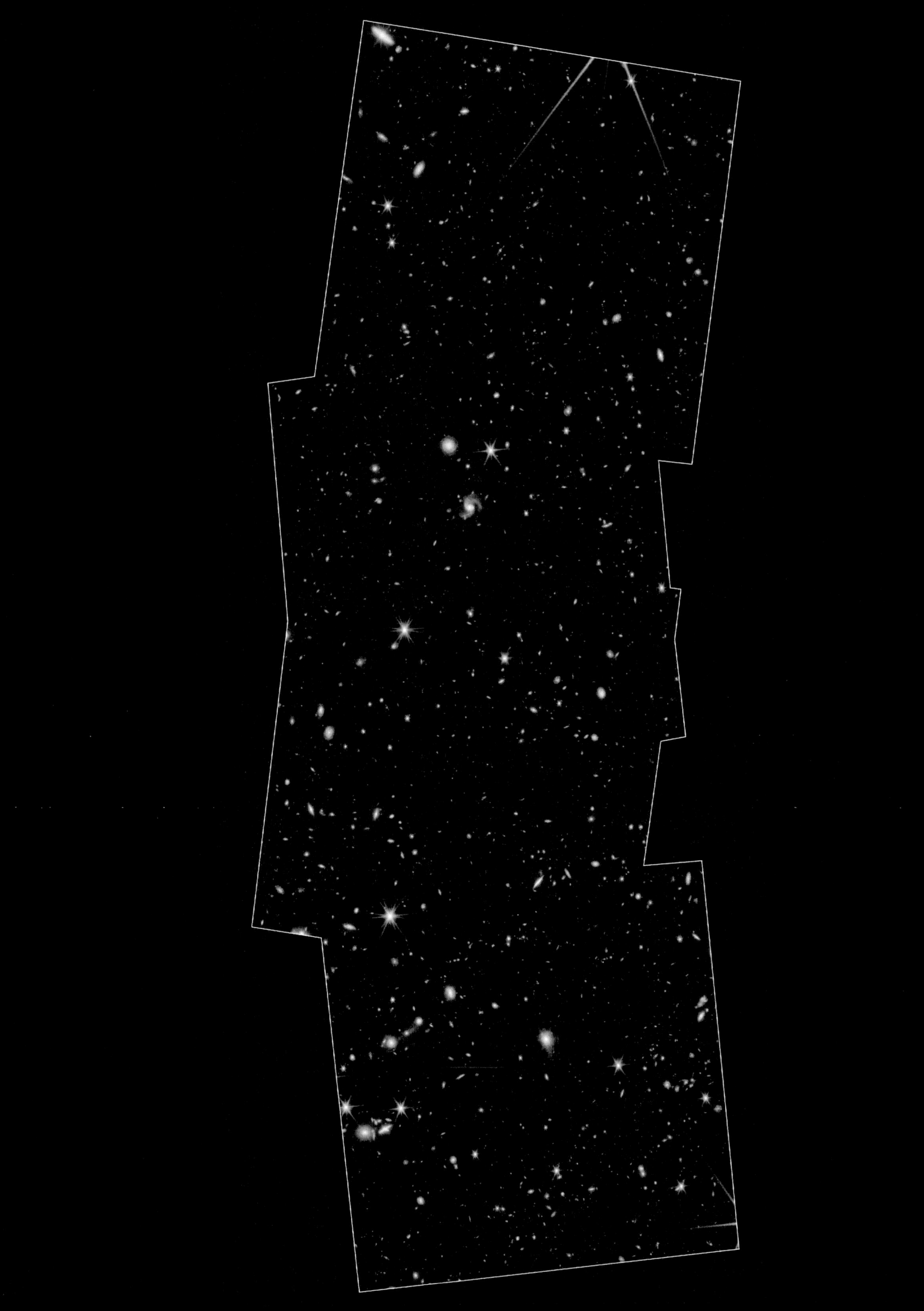

Jupiter Showcases Auroras, Hazes (NIRCam Closeup)

With giant storms, powerful winds, auroras, and extreme temperature and pressure conditions, Jupiter has a lot going on. Now, the NASA/ESA/CSA James Webb Space Telescope has captured new images of the planet. Webb's Jupiter observations will give scientists even more clues to Jupiter's inner life.

This image comes from the observatory's Near-Infrared Camera (NIRCam), which has three specialized infrared filters that showcase details of the planet. Since infrared light is invisible to the human eye, the light has been mapped onto the visible spectrum. Generally, the longest wavelengths appear redder and the shortest wavelengths are shown as more blue. Scientists collaborated with citizen scientist Judy Schmidt to translate the Webb data into images.

This image was created from a composite of several images from Webb. Visible auroras extend to high altitudes above both the northern and southern poles of Jupiter. The auroras shine in a filter that is mapped to redder colors, which also highlights light reflected from lower clouds and upper hazes. A different filter, mapped to yellows and greens, shows hazes swirling around the northern and southern poles. A third filter, mapped to blues, showcases light that is reflected from a deeper main cloud. The Great Red Spot, a famous storm so big it could swallow Earth, appears white in these views, as do other clouds, because they are reflecting a lot of sunlight.

Credit:NASA, ESA, Jupiter ERS Team; image processing by Judy Schmidt

Webb's First Deep Field (NIRCam Image)

housands of galaxies flood this near-infrared image of galaxy cluster SMACS 0723. High-resolution imaging from the NASA/ESA/CSA James Webb Space Telescope combined with a natural effect known as gravitational lensing made this finely detailed image possible.

First, focus on the galaxies responsible for the lensing: the bright white elliptical galaxy at the centre of the image and smaller white galaxies throughout the image. Bound together by gravity in a galaxy cluster, they are bending the light from galaxies that appear in the vast distances behind them. The combined mass of the galaxies and dark matter act as a cosmic telescope, creating magnified, contorted, and sometimes mirrored images of individual galaxies.

Clear examples of mirroring are found in the prominent orange arcs to the left and right of the brightest cluster galaxy. These are lensed galaxies – each individual galaxy is shown twice in one arc. Webb's image has fully revealed their bright cores, which are filled with stars, along with orange star clusters along their edges.

Not all galaxies in this field are mirrored – some are stretched. Others appear scattered by interactions with other galaxies, leaving trails of stars behind them.

Webb has refined the level of detail we can observe throughout this field. Very diffuse galaxies appear like collections of loosely bound dandelion seeds aloft in a breeze. Individual "pods" of star formation practically bloom within some of the most distant galaxies – the clearest, most detailed views of star clusters in the early universe so far.

One galaxy speckled with star clusters appears near the bottom end of the bright central star's vertical diffraction spike – just to the right of a long orange arc. The long, thin ladybug-like galaxy is flecked with pockets of star formation. Draw a line between its "wings" to roughly match up its star clusters, mirrored top to bottom. Because this galaxy is so magnified and its individual star clusters are so crisp, researchers will be able to study it in exquisite detail, which wasn't previously possible for galaxies this distant.

The galaxies in this scene that are farthest away – the tiniest galaxies that are located well behind the cluster – look nothing like the spiral and elliptical galaxies observed in the local universe. They are much clumpier and more irregular. Webb's highly detailed image may help researchers measure the ages and masses of star clusters within these distant galaxies. This might lead to more accurate models of galaxies that existed at cosmic "spring," when galaxies were sprouting tiny "buds" of new growth, actively interacting and merging, and had yet to develop into larger spirals. Ultimately, Webb's upcoming observations will help astronomers better understand how galaxies form and grow in the early universe.

Credit:NASA, ESA, CSA, and STScI

Webb Surprises Astronomers with Never-Before-Seen Details of the Early Universe

The massive gravity of galaxy cluster MACS0647 acts as a cosmic lens to bend and magnify light from the more distant MACS0647-JD system. It also triply lensed the JD system, causing its image to appear in three separate locations. These images, which are highlighted with white boxes, are marked JD1, JD2, and JD3. MACS0647-JD has a redshift of about 11, which puts it in the first 400 million years after the Big Bang. The long, diagonal line The long, diagonal line traversing the image is a diffraction spike from a bright star located just off the frame.

Credit:NASA, ESA, CSA, and STScI, Alyssa Pagan (STScI)

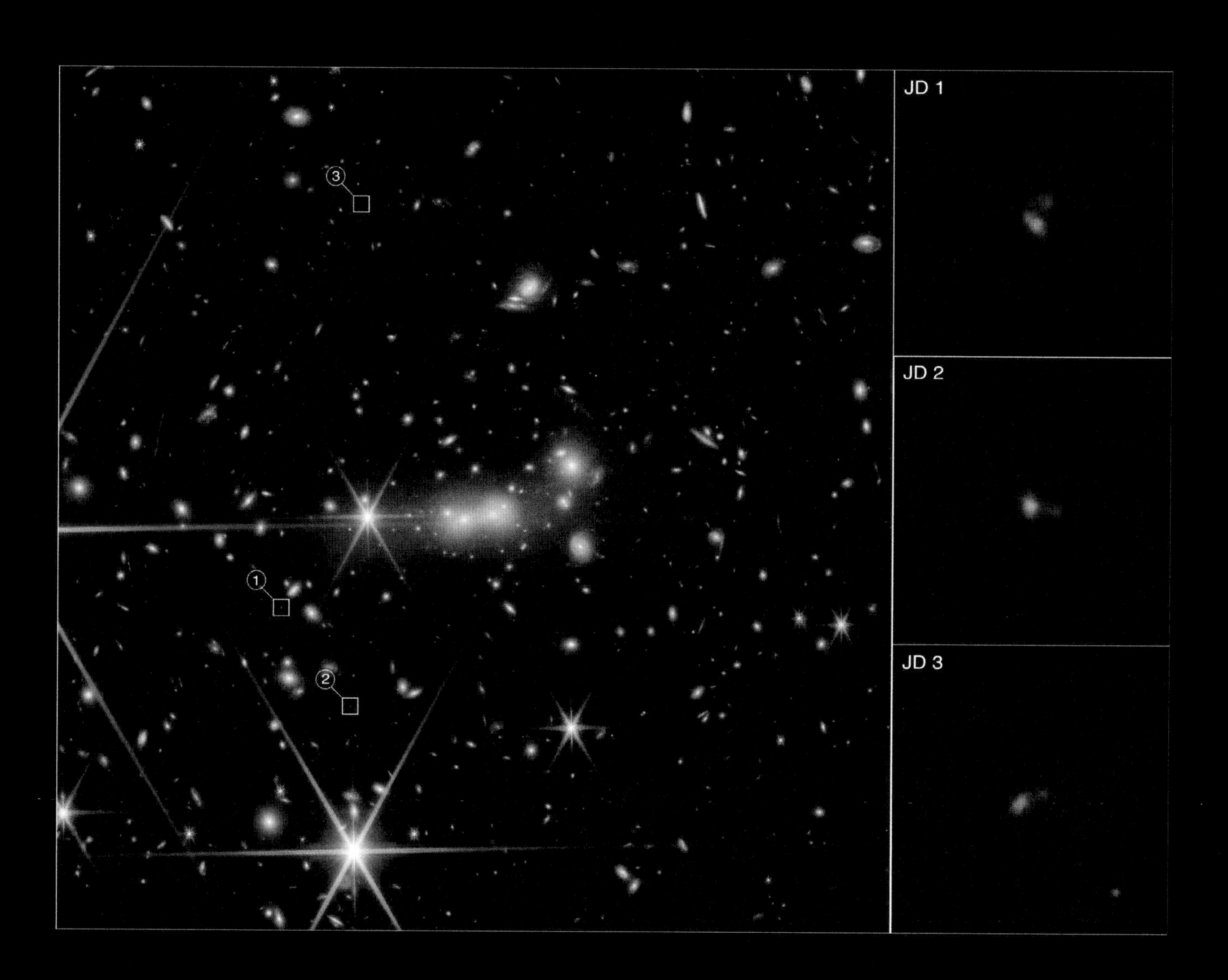
JD 1
JD 2
JD 3
1
2
3

Webb Finds Star Duo Forms 'Fingerprint' in Space

A new image from the NASA/ESA/CSA James Webb Space Telescope reveals a remarkable cosmic sight: at least 17 concentric dust rings emanating from a pair of stars. Located just over 5,000 light-years from Earth, the duo is collectively known as Wolf-Rayet 140.

Each ring was created when the two stars came close together and their stellar winds (streams of gas they blow into space) met, compressing the gas and forming dust. The stars' orbits bring them together about once every eight years; like the rings of a tree's trunk, the dust loops mark the passage of time.

In addition to Webb's overall sensitivity, its Mid-Infrared Instrument (MIRI) is uniquely qualified to study the dust rings, what Ryan Lau, the lead author from NSF's NOIRLab, and his colleagues call shells, because they are thicker and wider than they appear in the image. Webb's science instruments detect infrared light, a range of wavelengths invisible to the human eye.

MIRI detects the longest infrared wavelengths, which means it can often see cooler objects – including the dust rings – than Webb's other instruments can. MIRI's spectrometer also revealed the composition of the dust, formed mostly from material ejected by a type of star known as a Wolf-Rayet star.

A Wolf-Rayet star is born with at least 25 times more mass than our Sun and is nearing the end of its life, when it will likely collapse directly to black hole, or explode as a supernova. Burning hotter than in its youth, a Wolf-Rayet star generates powerful winds that push huge amounts of gas into space. The Wolf-Rayet star in this particular pair may have shed more than half its original mass via this process.

Credit:NASA, ESA, CSA, STScI, JPL-Caltech

Cartwheel Galaxy (NIRCam and MIRI Composite Image)

This image of the Cartwheel and its companion galaxies is a composite from Webb's Near-Infrared Camera (NIRCam) and Mid-Infrared Instrument (MIRI), which reveals details that are difficult to see in the individual images alone.

This galaxy formed as the result of a high-speed collision that occurred about 400 million years ago. The Cartwheel is composed of two rings, a bright inner ring and a colourful outer ring. Both rings expand outward from the centre of the collision like shockwaves.

However, despite the impact, much of the character of the large, spiral galaxy that existed before the collision remains, including its rotating arms. This leads to the "spokes" that inspired the name of the Cartwheel Galaxy, which are the bright red streaks seen between the inner and outer rings. These brilliant red hues, located not only throughout the Cartwheel, but also the companion spiral galaxy at the top left, are caused by glowing, hydrocarbon-rich dust.

In this near- and mid-infrared composite image, MIRI data are coloured red while NIRCam data are coloured blue, orange, and yellow. Amidst the red swirls of dust, there are many individual blue dots, which represent individual stars or pockets of star formation. NIRCam also defines the difference between the older star populations and dense dust in the core and the younger star populations outside of it.

Webb's observations capture Cartwheel in a very transitory stage. The form that the Cartwheel Galaxy will eventually take, given these two competing forces, is still a mystery. However, this snapshot provides perspective on what happened to the galaxy in the past and what it will do in the future.

Credit:NASA, ESA, CSA, STScI

Tarantula Nebula (NIRCam Image)

In this mosaic image stretching 340 light-years across, Webb's Near-Infrared Camera (NIRCam) displays the Tarantula Nebula star-forming region in a new light, including tens of thousands of never-before-seen young stars that were previously shrouded in cosmic dust. The most active region appears to sparkle with massive young stars, appearing pale blue. Scattered among them are still-embedded stars, appearing red, yet to emerge from the dusty cocoon of the nebula. NIRCam is able to detect these dust-enshrouded stars thanks to its unprecedented resolution at near-infrared wavelengths.

To the upper left of the cluster of young stars, and the top of the nebula's cavity, an older star prominently displays NIRCam's distinctive eight diffraction spikes, an artefact of the telescope's structure. Following the top central spike of this star upward, it almost points to a distinctive bubble in the cloud. Young stars still surrounded by dusty material are blowing this bubble, beginning to carve out their own cavity. Astronomers used two of Webb's spectrographs to take a closer look at this region and determine the chemical makeup of the star and its surrounding gas. This spectral information will tell astronomers about the age of the nebula and how many generations of star birth it has seen.

Farther from the core region of hot young stars, cooler gas takes on a rust colour, telling astronomers that the nebula is rich with complex hydrocarbons. This dense gas is the material that will form future stars. As winds from the massive stars sweep away gas and dust, some of it will pile up and, with gravity's help, form new stars.

Credit:NASA, ESA, CSA, and STScI

Neptune (NIRCam Image)

In this version of Webb's Near-Infrared Camera (NIRCam) image of Neptune, the planet's visible moons are labeled. Neptune has 14 known satellites, and seven of them are visible in this image.
Triton, the bright spot of light in the upper left of this image, far outshines Neptune because the planet's atmosphere is darkened by methane absorption wavelengths captured by Webb.

Triton reflects an average of 70 percent of the sunlight that hits it. Triton, which orbits Neptune in a backward orbit, is suspected to have originally been a Kuiper belt object that was gravitationally captured by Neptune.

Credit:NASA, ESA, CSA, and STScI

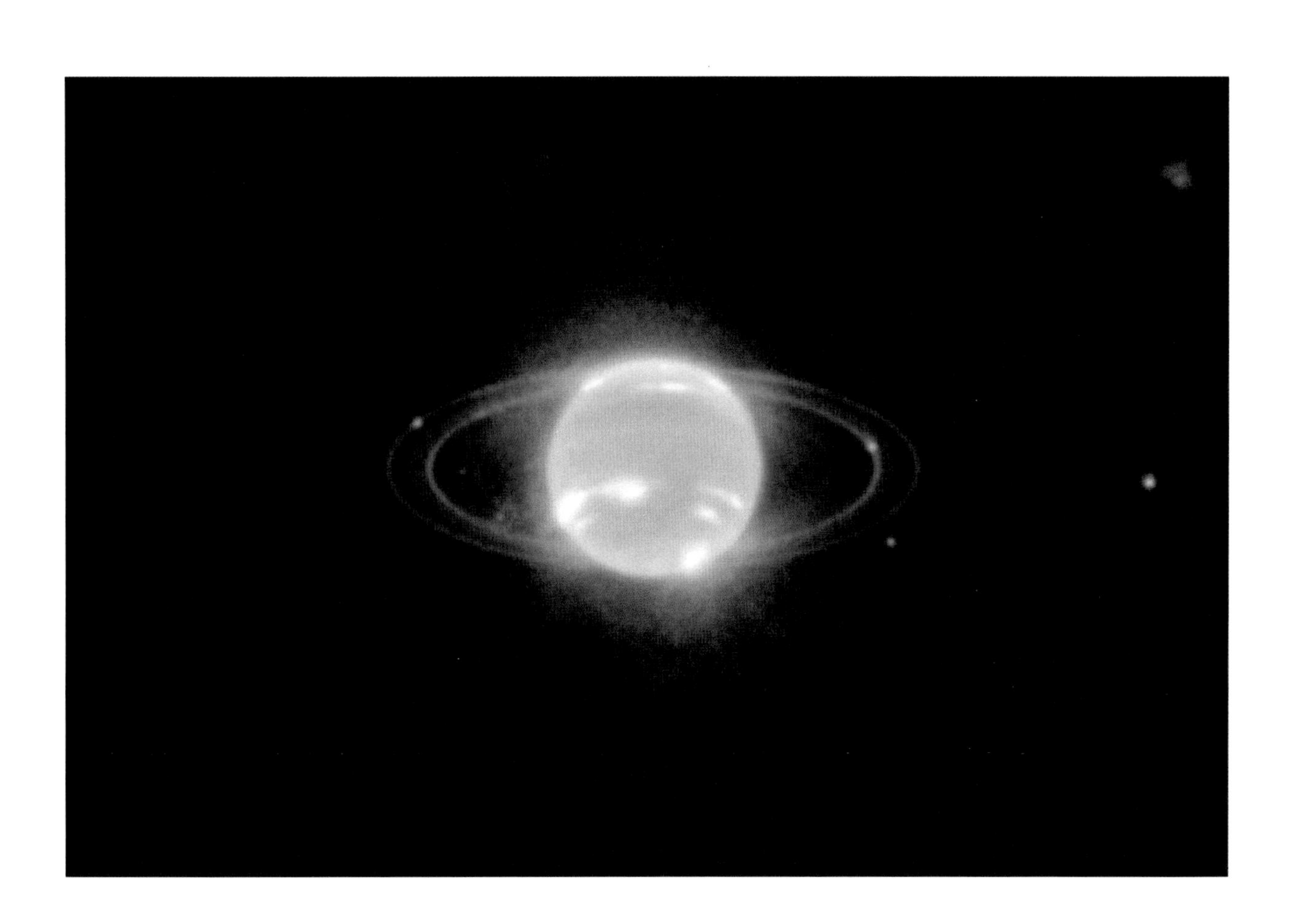

Stephan's Quintet

Webb's Mid-InfraRed Instrument (MIRI) shows never-before-seen details of Stephan's Quintet, a visual grouping of five galaxies, in this image. MIRI pierced through dust-enshrouded regions to reveal huge shock waves and tidal tails, gas and stars stripped from the outer regions of the galaxies by interactions. It also unveiled hidden areas of star formation. The new information from MIRI provides invaluable insights into how galactic interactions may have driven galaxy evolution in the early universe. This image was released as part of the first set of images from the NASA/ESA/CSA James Webb Space Telescope on 12 July 2022.

This image contains one more MIRI filter than was used in the NIRCam-MIRI composite picture. The image processing specialists at the Space Telescope Science Institute in Baltimore opted to use all three MIRI filters and the colours red, green and blue to most clearly differentiate the galaxy features from each other and the shock waves between the galaxies.

In this image, red denotes dusty, star-forming regions, as well as extremely distant, early galaxies and galaxies enshrouded in thick dust. Blue point sources show stars or star clusters without dust. Diffuse areas of blue indicate dust that has a significant amount of large hydrocarbon molecules. For small background galaxies scattered throughout the image, the green and yellow colours represent more distant, earlier galaxies that are rich in these hydrocarbons as well.

Stephan's Quintet's topmost galaxy – NGC 7319 – harbours a supermassive black hole 24 million times the mass of the Sun. It is actively accreting material and puts out light energy equivalent to 40 billion Suns. MIRI sees through the dust surrounding this black hole to unveil the strikingly bright active galactic nucleus.

As a bonus, the deep mid-infrared sensitivity of MIRI revealed a sea of previously unresolved background galaxies reminiscent of Hubble's Deep Fields.

Together, the five galaxies of Stephan's Quintet are also known as the Hickson Compact Group 92 (HCG 92). Although called a "quintet," only four of the galaxies are truly close together and caught up in a cosmic dance. The fifth and leftmost galaxy, called NGC 7320, is well in the foreground compared with the other four. NGC 7320 resides 40 million light-years from Earth, while the other four galaxies (NGC 7317, NGC 7318A, NGC 7318B, and NGC 7319) are about 290 million light-years away. This is still fairly close in cosmic terms, compared with more distant galaxies billions of light-years away. Studying these relatively nearby galaxies helps scientists better understand structures seen in a much more distant universe.

This proximity provides astronomers a ringside seat for witnessing the merging of and interactions between galaxies that are so crucial to all of galaxy evolution. Rarely do scientists see in so much exquisite detail how interacting galaxies trigger star formation in each other, and how the gas in these galaxies is being disturbed. Stephan's Quintet is a fantastic "laboratory" for studying these processes fundamental to all galaxies.
Tight groups like this may have been more common in the early universe when their superheated, infalling material may have fueled very energetic black holes called quasars. Even today, the topmost galaxy in the group – NGC 7319 – harbours an active galactic nucleus, a supermassive black hole that is actively pulling in material.

Credit:NASA, ESA, CSA, and STScI

Webb's First Deep Field (MIRI and NIRCam Images Side by Side)

Galaxy cluster SMACS 0723 is a technicolour landscape when viewed in mid-infrared light by the NASA/ESA/CSA James Webb Space Telescope. Compared to Webb's near-infrared image at right, the galaxies and stars are awash in new colours.

Start by comparing the largest bright blue star. At right, it has very long diffraction spikes, but in mid-infrared at left, its smaller points appear more like a snowflake's. Find more stars by looking for these telltale – if tiny – spikes. Stars also appear yellow, sometimes with green diffraction spikes.
If an object is blue and lacks spikes, it's a galaxy. These galaxies contain stars, but very little dust. This means that their stars are older – there is less gas and dust available to condense to form new stars. It also means their stars are ageing.

The red objects in this field are enshrouded in thick layers of dust, and may very well be distant galaxies. Some may be stars, but research is needed to fully identify each object in the mid-infrared image.

The prominent arcs at the centre of the galaxy cluster, which are galaxies that are stretched and magnified by gravitational lensing, appear blue in the Mid-Infrared Instrument (MIRI) image at left and orange in the Near-Infrared Camera (NIRCam) image at right. These galaxies are older and have much less dust.

Galaxies' sizes in both images offer clues as to how distant they may be – the smaller the object, the more distant it is. In mid-infrared light, galaxies that are closer appear whiter.

Among this kaleidoscope of colours in the MIRI image, green is the most tantalising. Green indicates a galaxy's dust includes a mix of hydrocarbons and other chemical compounds.

The differences in Webb's images are owed to the technical capabilities of the MIRI and NIRCam instruments. MIRI captures mid-infrared light, which highlights where the dust is. Dust is a major ingredient for star formation. Stars are brighter at shorter wavelengths, which is why they appear with prominent diffraction spikes in the NIRCam image.

With Webb's mid-infrared data, researchers will soon be able to add much more precise calculations of dust quantities in stars and galaxies to their models, and begin to more clearly understand how galaxies at any distance form and change over time.

Credit:ESA/CNES/Arianespace/Optique Vidéo du CSG - JM Guillon

Webb's First Image of Focused Star

On 11 March 2022, the NASA/ESA/CSA James Webb Space Telescope (Webb) team completed the stage of mirror alignment known as "fine phasing". Although there are months to go before Webb ultimately delivers its new view of the cosmos, achieving this milestone means the team is confident that Webb's first-of-its-kind optical system is working as well as possible.

At this key stage in the commissioning of Webb's Optical Telescope Element, every optical parameter that has been checked and tested is performing at, or above, expectations. The team also found no critical issues and no measurable contamination or blockages to Webb's optical path: the observatory is able to successfully gather light from distant objects and deliver it to its instruments without issue. At this stage of Webb's mirror alignment, each of the primary mirror segments has been adjusted to produce one unified image of the same star using only Webb's primary imager, the Near-Infrared Camera (NIRCam), and NIRCam has been fully aligned to the observatory's mirrors.

While the purpose of this image was to focus on the bright star at the centre (called 2MASS J17554042+6551277) for alignment evaluation, Webb's optics and NIRCam are so sensitive that galaxies and stars in the background also show up. This image uses a red filter to optimise visual contrast.

Credit:NASA/STScI

TELESCOPE ALIGNMENT EVALUATION IMAGE

Fine Guidance Sensor Test Image

Webb's Fine Guidance Sensor (FGS) – developed by the Canadian Space Agency – has captured a special view of stars and galaxies that provides a tantalising glimpse at what the telescope's science instruments will reveal in the coming weeks, months, and years.

The FGS has always been capable of capturing imagery, but its primary purpose is to enable accurate science measurements and imaging with precision pointing. When it does capture imagery, the imagery is typically not kept: Given the limited communications bandwidth between L2 and Earth, Webb only sends data from up to two science instruments at a time. But during a week-long stability test in May, it occurred to the team that they could keep the imagery that was being captured because there was available data transfer bandwidth.

The resulting engineering test image has some rough-around-the-edges qualities to it. It was not optimised to be a science observation; rather, the data was taken to test how well the telescope could stay locked onto a target, but it does hint at the power of the telescope. It carries a few hallmarks of the views Webb has produced during its post launch preparations. Bright stars stand out with their six, long, sharply defined diffraction spikes – an effect due to Webb's six-sided mirror segments. Beyond the stars, galaxies fill nearly the entire background.

The result – using 72 exposures over 32 hours – is among the deepest images of the universe ever taken, according to Webb scientists. When FGS' aperture is open, it is not using colour filters like the other science instruments – meaning it is impossible to study the age of the galaxies in this image with the rigour needed for scientific analysis. But even when capturing unplanned imagery during a test, FGS is capable of producing stunning views of the cosmos.

Credit:NASA, CSA, and FGS team.

MIRI and Spitzer Comparison Image

The NASA/ESA/CSA James Webb Space Telescope is aligned across all four of its science instruments. Here we take a closer look at Webb's coldest instrument: the Mid-Infrared Instrument, or MIRI.

This MIRI test image (at 7.7 microns) shows part of the Large Magellanic Cloud. This small satellite galaxy of the Milky Way provided a dense star field to test Webb's performance.

Here, a close-up of the MIRI image is compared to a past image of the same target taken with NASA's Spitzer Space Telescope's Infrared Array Camera (at 8.0 microns). The retired Spitzer was the first observatory to provide high-resolution images of the near- and mid-infrared Universe. Webb, by virtue of its significantly larger primary mirror and improved detectors, will allow us to see the infrared sky with improved clarity, enabling even more discoveries.

For example, Webb's MIRI image shows the interstellar gas in unprecedented detail. Here, you can see the emission from 'polycyclic aromatic hydrocarbons' – molecules of carbon and hydrogen that play an important role in the thermal balance and chemistry of interstellar gas. When Webb is ready to begin science observations, studies such as these with MIRI will help give astronomers new insights into the birth of stars and protoplanetary systems.

Credit:NASA/JPL-Caltech; MIRI: NASA/ESA/CSA/STScI

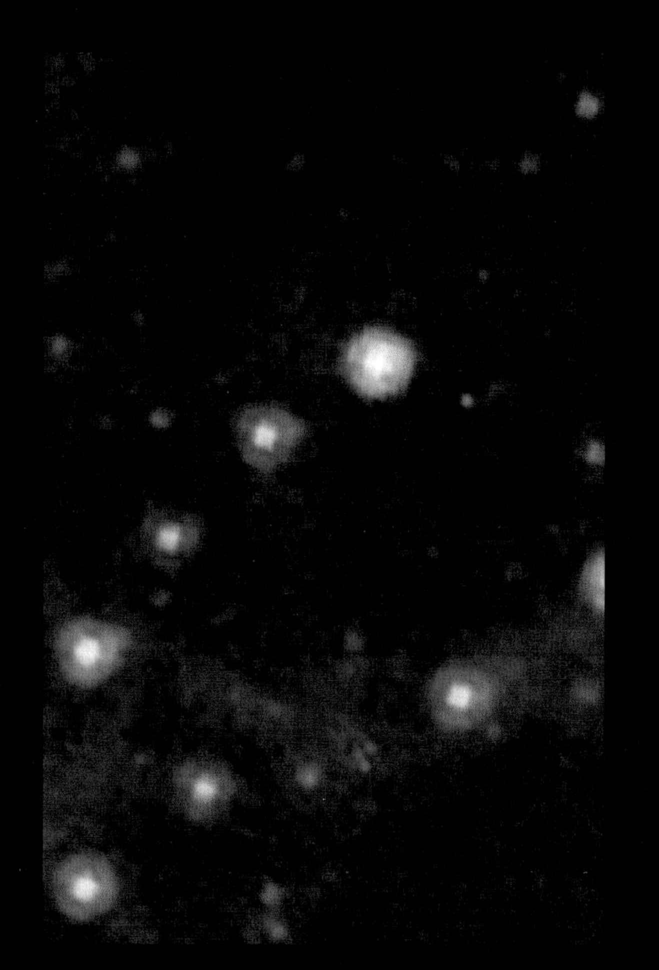

SPITZER IRAC 8.0μ

WEBB MIRI 7.7μ

Webb launch celebration

ESA (European Space Agency) Director-General Dr. Josef Aschbacher, 2nd from left, and NASA Associate Administrator for the Science Mission Directorate Thomas Zurbuchen, 3rd from left, celebrate after hearing confirmation that the James Webb Space Telescope successfully separated from the Ariane 5 rocket, Saturday, Dec. 25, 2021, in the Jupiter Hall of the Guiana Space Centre in Kourou, French Guiana. The James Webb Space Telescope (sometimes called JWST or Webb) is a large infrared telescope with a 21.3 foot (6.5 meter) primary mirror. The observatory will study every phase of cosmic history—from within our solar system to the most distant observable galaxies in the early universe.

Credit:NASA/Bill Ingalls

arianespace
WEBB
LAUNCH

Webb Liftoff on Ariane 5

The NASA/ESA/CSA James Webb Space Telescope lifted off on an Ariane 5 rocket from Europe's Spaceport in French Guiana, at 13:20 CET on 25 December on its exciting mission to unlock the secrets of the Universe.

Following launch and separation from the rocket, Webb's mission operations centre in Baltimore, USA confirmed Webb deployed its solar array and is in good condition, marking the launch a success.

In the coming month, Webb, an international partnership between NASA, ESA and the Canadian Space Agency (CSA), will travel to its destination: the second Lagrange point (L2), where it will study the Universe in infrared.

Credit:ESA/CNES/Arianespace/Optique Vidéo du CSG - JM Guillon

esa
ariane

Webb Mirror Closeup

The NASA/ESA/CSA James Webb Space Telescope is shown here with its primary mirror fully deployed during a Comprehensive Systems Test in 2020 at Northrop Grumman in California, USA.

This was the first full systems evaluation that was run on the assembled observatory.

Credit:NASA/Chris Gunn

*“We are ourselves composed of the dust of the Big Bang.
Perhaps we carry within us the memory of the universe?”*

Hubert Reeves

Unveiling the Cosmos: Your Journey Continues

As you near the end of this captivating book, you've journeyed through the spellbinding beauty and immense expanse of the universe, all captured through the lens of the James Webb Space Telescope. Each turned page has been a portal to distant realms and celestial spectacles that stretch the boundaries of human comprehension. But let me share a secret with you – the adventure has only just begun. Imagine diving into uncharted cosmic territories, where unexplored galaxies await your gaze, where the intricate dance of starbirth and destruction unfolds in breathtaking detail, and where the birthplaces of planets are unveiled in unprecedented clarity.

As you close this chapter, don't close the door on discovery. The forthcoming volumes in this series promise an odyssey beyond your wildest imagination. Picture immersing yourself in the vibrant landscapes of nebulae, witnessing the birth and evolution of stars like never before, and gaining unprecedented insights into the origins of the universe itself. With each volume, you'll be transported deeper into the heart of space, guided by the keen eyes of the James Webb Space Telescope.

So, don't miss out on the continuation of this cosmic journey. Embrace the opportunity to venture further, to learn more, and to stand in awe of the majesty of the cosmos. The upcoming installments are poised to elevate your understanding and wonder to even greater heights, building upon the foundation laid by the book in your hands. Get ready to explore, to marvel, and to unlock the secrets that the universe eagerly awaits to share.

Made in the USA
Monee, IL
20 December 2023

ae58c94b-8113-4012-a8be-0d3d2c9a4aadR02